D0102729

Contents

Preface v

1 Chapter 1 1

A - Measurement 1

A-1 Reviewing data 1

A-2 Continuous Measurement Procedures 2

A-3 Discontinuous Measurement 7

A-4 - Permanent Product 12

A-5 - Data and Graphs 13

A-6 - Describe Behavior and the Environment 16

2 Chapter 2 18

B - Assessment 18

B-1 Preference Assessment 18

B-2 Functional Assessment Procedures 24

B-3 Functional Behavior Assessment 25

3 Chapter 3 30

C -Skills Acquisition 30

C-1 Skill Acquisition Plan 30

C-2 Session Preparation 32

C-3 Contingencies of Reinforcement 33

Conditioned vs. Unconditioned 33

C-4 Discrete Trial Teaching 37

C-5 Naturalistic Teaching Procedures 38

C-6 Task Analysis 39

C-7 Discrimination Training 40

C-8 Stimulus Control Transfer 41

C-9 Prompt and Prompt Fading 42
C-10 Generalization and Maintenance 44
C-11 Shaping Procedures 45
C-12 Token Economy 46
4 Chapter 4 48
D - Behavior Reduction 48
D-1 Identify the Essential Components of a Behavior Reduction Plan 49
D-2 Functions of Behavior 49
D-3 Interventions based on Antecedents 50
D-4 Differential Reinforcement 51
D-5 Extinction Procedures 52
D-6 Crisis/Emergency Procedures 53
5 Chapter 5 55
E - Documentation and Reporting 55
E-1 - Effectively Communicate 55
E-2 - Seek Clinical Direction 56
E-3 - Report on Variables 57
E-4 - Objective Session Notes 57
E-5 - Comply with Applicable Legal Requirements 58
6 Chapter 6 59
F - Professional Conduct and Scope of Practice 59
F-1 - Describe the Role of RBT 59
F-2 - Respond Appropriately to Feedback 62
F-3 - Communicate with Stakeholders 63
F-4 - Maintain Professional Boundaries 65
F-5 - Maintain Client Dignity 68
7 Conclusion 70
About the Author 71

Preface

This study guide provided content information on each task in the RBT® Task List 2nd Edition. ®BCBA, RBT, BACB, or any other BACB trademarks used is/are registered to the Behavior Analyst Certification Board® ("BACB®"). Rachel White is not sponsored, endorsed, or affiliated with the BACB® other than the BCBA certification.

This study guide is for educational purposes and is not a guarantee of passing the BACB certification exams. We do not guarantee refunds if you fail your exam. We are not affiliated with the bacb or the exam in any way. Our questions and content are based off the cooper book, and not what you would find on the exam verbatim.

I have been tutoring and training RBTs for a few years, and I find that everyone wants to memorize the content for the exam; however, they have difficulty applying the concepts. This lack of application knowledge overwhelms RBTs when they begin working with clients. The feeling of being overwhelmed and limited knowledge on appropriately addressing behaviors and behavior goals leads to RBT burnout and turnover.

The goal of this study guide is to provide a solid foundation. With this solid foundation, the RBT can learn to apply the concepts while working with the clients.

Keep a look out for more study materials! You can also locate the latest information on the behavior.prep.com website.

1

Chapter 1

A - Measurement

A-1 Reviewing data

Prepare for data collection

Review data from the last session

Determine targets to be run in the upcoming session based on the data review

Gather materials needed for the upcoming session

Set up for the session

Data collection is a critical job aspect for an RBT. Data collection is used to make program decisions for each client. Missing data or data collection errors could result in uninformed program changes, which ultimately could cause harm to your client. Make sure you are always knowledgeable about your client's data for each target in the program and set up a session environment that results in reliable, accurate data collection. Some companies use digital or online data collection tools, while others may only use paper to collect data. As an RBT, you must be aware of all data collection methods that can be used at your place of employment.

A-2 Continuous Measurement Procedures

Implement continuous measurement procedures

Frequency

Rate

Duration

Latency

Interresponse time

Continuous measurement procedures are used to measure all instances of a behavior. The RBT continuously watches the client and records every time the behavior occurs.

The following table provides an overview of continuous measurement procedures.

Procedure	Definition	Example
Frequency	A count for each time the behavior occurs. Frequency is best to use when the observation length (e.g. therapy session) is consistently the same.	Robin hit her head 4 times. The frequency is 4. The RBT "counts" how many times Robin hit her head.
Rate	Ratio of count per observation period. Not to be confused with frequency. It is expressed as the frequency divided by the time period. Rate is usually used when the observation length is not consistently the same.	Robin hit her head 4 times the first hour and 2 times the second hour. The rate is 3 times per hour.
Duration	Total length of time a behavior occurs	Robin screamed for 4 minutes when she was denied access to her toy.
Latency	The amount of time from the onset of the stimulus to the start of the response	The teacher handed out a worksheet facedown. She instructed the students to turn the paper over and begin working. Charlie waited 15 seconds before he turned the paper over to work. The latency is 15 seconds.
Interresponse Time	The amount of time between responses	A personal trainer is working with Sue. The trainer instructs Sue to do 10 push-ups. Sue waits 10 seconds between push-ups. The interresponse time is 10 seconds.

Frequency vs. Rate Data Collection

Frequency and rate both count the occurrence of behavior; however, the data are calculated differently. Rate includes frequency as well as the time period. The following table provides an overview of collecting data for each procedure.

Date	Client:	Target Behavior	Observation Interval
4/2/2020	ML	Tapping index finger on table	Interval: 2 minutes Total time: 10 minutes Number of trials: 3

	1	2	3	4	5	Total
Trial 1	5	4	8	5	7	29
Trial 2	7	5	8	9	6	35
Trial 3	4	4	9	7	4	25
		Total for 10-minute interval			**89**	

The data in the chart above show the number of times (frequency) the client tapped her index finger on the table during each interval. The client tapped her finger five times in the first interval of Trial 1. At the end of Trial 1, the client tapped her finger 29 times (frequency) in 10 minutes, or 2.9 taps/min (rate). For the entire 30-minute duration, the client tapped her finger 89 times (frequency), or 2.96 taps/minute (rate).

Duration

Duration is another continuous measurement procedure used to measure how long a behavior occurs. Duration records instances of behavior that occur too frequently to count or when a behavior is continuous. The definition of the behavior determines how the data are recorded. The following table provides an overview of recording duration data.

Duration Example	
Target Behavior	Wear mask over mouth and nose without exhibiting escape behavior in the form of tantrum behavior (crying, screaming, stomping feet)

Behavior Definition	Duration data	Rationale
Marge will wear her mask for 1 minute without attempting to remove the mask	27 seconds	The behavior is recorded as: Mask on for 10 seconds, off 15 seconds, mask on for 24 seconds. The total duration for the behavior is 49
	31 seconds	The behavior is recorded as: 10 seconds tap, 20 second pause, 15 seconds tap, 4 seconds' pause. The duration for the behavior is 31 because there was no pause more than 5 seconds

Duration provides more information about the occurrence of a behavior that is more than a few seconds long. For example, you may take frequency data on tantrums. If the client exhibits 2 tantrums during a 2-hour session, this information does not provide an accurate account of the behavior. Adding duration offers more information. Use the same example of 2 tantrums during a 2-hour session. When duration is used in conjunction with frequency, the first tantrum lasting 15 minutes and the second tantrum lasting 35 minutes is very different from the first tantrum lasting 45 seconds and the second tantrum lasting 25 seconds.

Latency

Latency is used to determine the amount of time between the presentation of a stimulus and a response. Latency can be used to increase or decrease behavior.

Interresponse Time (IRT)

Interresponse time is used to determine the amount of time between the presentation of two consecutive responses. Interresponse can be used to increase or decrease behavior. The following table provides an overview of collecting interresponse time data.

Interresponse Time Example		
Data Collection	**IRT**	**Rationale**
Sara puts food in her mouth before she has completed chewing and swallowing what is already in her mouth. This behavior results in her mouth being overfull, as well as several instances of choking. The BCBA determines that Sara puts food in her mouth on an average of every 10 seconds.	10 seconds	Sara's IRT for taking bites is very short and results in safety issues. To address these safety concerns the IRT between bites is increased to allow more time to chew and swallow.

When to Use Continuous Measurement

Procedure	Use when:
Frequency	• The observation length is consistent from day to day • The behavior is discrete and short in duration
Rate	• The observation length varies from day to day • The behavior lasts for more than a few seconds
Duration	• There is a clear beginning and end to the behavior • The length of the behavior is a concern • Behavior occurs at a high frequency
Latency	• There is a clear beginning and end to the behavior • It is important to know the "time to respond"
Interresponse Time	• There is a clear pause between two consecutive responses • Determining if a behavior should be counted as more than one instance

A-3 Discontinuous Measurement

Implement discontinuous measurement procedures

Partial interval

Whole interval

Momentary time sampling

Discontinuous measurement procedures collect data on a behavior during a specified interval. Continuous measurement procedures record every occurrence of a behavior; discontinuous measures are used if a behavior occurs during or at the end of a specified interval. When using discontinuous measurement procedures, you estimate a behavior's occurrence rather than recording exact measurements. Data is recorded as either the behavior occurring or not occurring; it does not record the number of times the behavior occurs.

A stopwatch, clock, or wristwatch is needed to keep track of the time intervals.

The following table provides an overview of discontinuous measurement time sampling procedures.

Procedure	Definition	Example
Partial interval	A time sampling procedure in which a behavior is recorded if it occurs at least once in during the defined interval.	The RBT is taking data on picking fingers during a 5-minute interval. At the 3-minute mark, the client exhibits the behavior for 10 seconds. The interval is marked as the behavior occurring.
Whole interval	A time sampling procedure in which a behavior is recorded only if it occurs during the entire defined interval.	The RBT is taking data on picking fingers during a 5-minute interval. At the 3-minute mark, the client exhibits the behavior for 10 seconds. The interval is marked as the behavior not occurring because the behavior did not occur the entire interval.
Momentary time sampling	A time sampling procedure in which the behavior is recorded if it occurs at the end of a defined interval.	The RBT is taking data on picking fingers during a 5-minute interval. At the 3-minute mark, the client exhibits the behavior for 10 seconds. The interval is marked as the behavior not occurring because the behavior did not occur at the end of the interval.

Partial Interval

Partial interval recording measures behaviors that happen so quickly that they are hard to catch. A time period is broken into intervals. The RBT records if the behavior occurred at least once during the interval, and the data are entered as a percentage. The following table provides an overview of recording partial interval data.

Partial Interval Example			
Date	**Client:**	**Target Behavior**	**Observation Interval**
4/2/2022	ML	Spinning behavior	Interval: 2 minutes Total time: 10 minutes

1	2	3	4	5	Total
X	+	+	X	X	40%

In the table above, the behavior only occurred in intervals 2 and 3. The number of intervals in which the behaviors occurred is divided by the total number of intervals to determine the percentage. There are a total of 5 intervals. The behavior occurred in two intervals—⅖=40 %.

Partial interval recording tends to overestimate the occurrence of the behavior. Use the example from the table in which the behavior occurred 40% of the time. Based on the percentage, you would conclude that the behavior may have occurred for 4 minutes; however, this is not the case. When collecting partial interval data, the behavior is recorded as happening even if the behavior occurred for a short duration. In this example, the behavior could have occurred between 2 seconds - 4 minutes.

Whole Interval

The whole interval is used to measure behaviors that occur the entire interval. This procedure requires undivided attention for the RBT. A time period is broken into intervals. The RBT records if the behavior occurs during the entire interval, and the overall data are entered as a percentage. The following table provides an overview of recording whole interval data.

Whole Interval Example			
Date	**Client:**	**Target Behavior**	**Observation Interval:**
4/2/2022	ML	Humming behavior during seat work	Interval: 2 minutes Total time: 10 minutes

1	2	3	4	5	Total
+	+	X	X	+	60%

Whole interval recording tends to underestimate the occurrence of the behavior. Use the example from the table in which the behavior occurred 60% of the time. Based on the percentage, you would conclude that the behavior may have occurred for 6 minutes; however, this is not the case. When collecting whole interval data, the behavior is recorded as happening only if the behavior occurs during the entire interval. In this example, the behavior is marked as occurring during intervals 1,2 and 5. This data reflects that the behavior occurred for at least 6 minutes. Even if the behavior occurred for 30 seconds in interval 3 and 1 minute in interval 4, the intervals are marked as not occurring.

Momentary Time Sampling

Momentary time sampling measures behaviors that occur at the end of an interval. A time period is broken into intervals. The RBT records if the behavior occurs only at the end of the interval, and the overall data for the time period is entered as a percentage.

Momentary Time Sampling Example			
Date	**Client:**	**Target Behavior**	**Observation Interval:**
4/2/2022	ML	Humming behavior during seat work	Interval: 2 minutes Total time: 10 minutes

1	2	3	4	5	Total
X	X	X	+	+	40%

Momentary time sampling tends to underestimate and overestimate the occurrence of the behavior. Use the example from the table in which the behavior occurred 40% of the time. Based on the percentage, you would conclude that the behavior occurred for 4 minutes; however, this is not the case. When collecting momentary time sampling interval data, the behavior is recorded as happening only if the behavior occurred at the end of the interval. In this example, the behavior occurs during intervals 4 and 5. This data reflects that the behavior could have occurred for at least 4 minutes. Even if the behavior occurred at the beginning of the interval for 45 seconds, the interval is still marked as not occurring because it did not occur at the end of the interval. This scoring would underestimate the behavior. If the behavior occurred for only the last 5 seconds of interval 5, it is marked as occurring because it occurred at the end of the interval. This scoring overestimates the behavior.

When to Use Discontinuous Measurement

Procedure	Use when:
Partial interval	• The behavior happens quickly • The behavior does not last long • You want to decrease behavior • It is not possible to observe continuously
Whole interval	• The behavior is not easily counted • The behavior does not have a clear beginning or end • The behavior occurs at a high frequency • You want to increase a behavior
Momentary Time Sampling	• The behavior is not easily counted • The behavior does not have a clear beginning or end • It is not possible to observe continuously

A-4 - Permanent Product

Implement permanent product recording procedures

Permanent product

A permanent product is the result of an individual's behavior. A permanent product might be a written essay, a math test, making dinner, or washing clothes. Permanent products are examined to determine if a task has been completed. Direct observation of a behavior is not required when implementing permanent product procedures because the permanent product indicates that the behavior occurred.

The following table provides examples of using a permanent product procedure.

Behavior Task	Permanent Product
Using the microwave to make a snack	A microwave bag of popcorn
80% accuracy on math test	Grade of 80 or above on math test
Clear out the dishwasher	Empty dishwasher

A-5 - Data and Graphs

Enter data and update graphs

Level

Trend

Variability

Data for targets are gathered each session. The data are put into a graph to examine the client's progress visually. The success, failure, or stagnation of a target can be determined from the graph. Graphs are reviewed to determine the following:

- Level - the position of data on the Y axis
- Trend - the direction in which that data is moving
- Variability - the relationship between the data points on the graph

Level

Identify the lowest and highest data points to determine the level of a graph. Use the graph below to determine the lowest and highest data points.

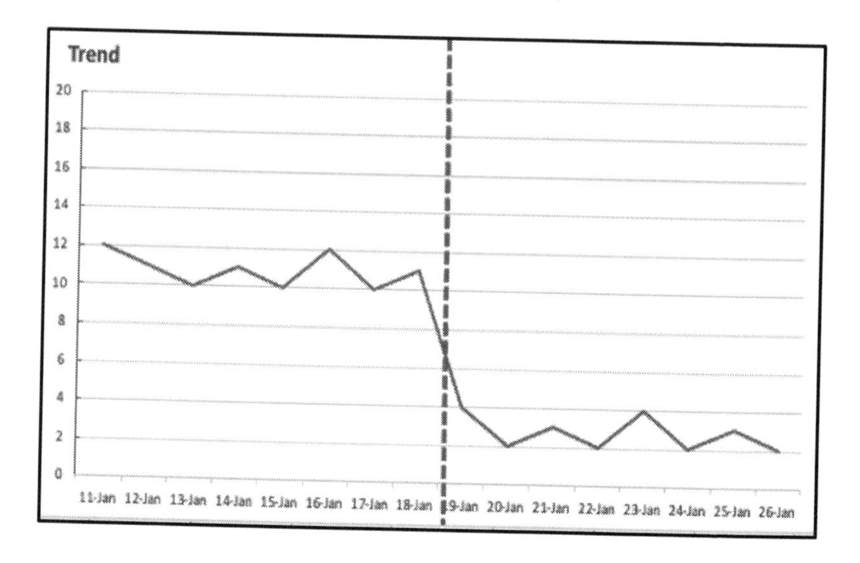

Look to the left of the red line. The data points to the left show a level of between 10-12. The data points to the right indicate a level between 2-4. There is a change in level between the two sides of the graph. The left side shows a level of high rates of responding, and the right side shows a level of low rates of responding.

Trend

Trend relates to the overall direction of the data on the graph. Trends can be increasing, decreasing, or no trend. The following graph provides a visual of the three types of trend lines.

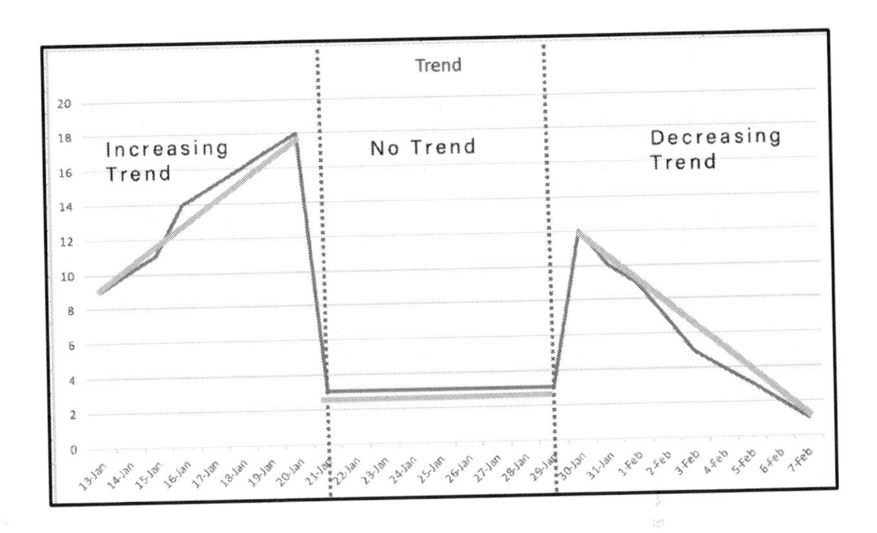

Variability

Variability is determined by comparing data points to other data points on the graph. Variability results in the inability to determine if an intervention is working because the data points do not show consistency in trend or level. The following graph displays variability.

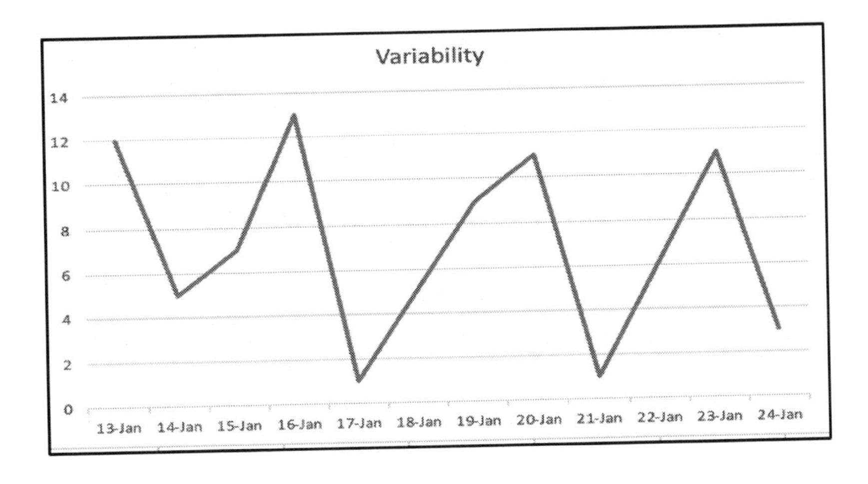

Data and graphs are critical to understanding and interrupt to make clinical decisions about a client's program. Graphs can be generated using tools such as Microsoft Excel, Google Sheets, or a third-party software provider.

A-6 - Describe Behavior and the Environment

Describe behavior and environment in observable and measurable terms.

Observable

Measurable

Observable and measurable terms are used to describe a behavior and its form. Defining the behavior in these terms allows a therapist to observe the target behavior. Once observed, the therapist can then measure the behavior by taking data.

The following table provides an example of a definition.

Example
Twilla slaps herself on the side of the head using an open palm as a function of escape when she is presented with non-preferred tasks.

In this example, the target behavior is slapping. It is defined as using an open palm and hitting the side of the head and only occurs with non-preferred tasks.

With this definition, a new therapist, in Twilla's case, can observe and measure this behavior.

Non-Example
Twilla hits herself when she is given work.

In this example, the target behavior is hitting; however, it is unknown if she uses a closed fist or open palm and where she is hitting herself. The antecedent is also unknown.

With this definition, a new therapist in Twilla's case would be unable to observe and measure this behavior.

Guidelines

When describing behaviors, make sure to use:

- Objective terms, not subjective or opinions such as bad behavior, mean behavior, or out of control
- Use the function of the behavior when possible
- Use the antecedent behavior, if known

2

Chapter 2

B - Assessment

B-1 Preference Assessment

Conduct preference assessments

Free operant

Single choice

Paired choice

Multiple with replacement

Multiple without replacement

Preference assessments are conducted to determine items that may be used as reinforcers during a session. The word "may" is used because you will not know if an item is reinforcing until you run the program. The following list provides an overview of preference assessments:

- Free operant

- Single stimulus
- Paired stimulus (Forced choice)
- Multiple stimulus with replacement
- Multiple stimulus without replacement

Free Operant Preference Assessment

In a free operant preference assessment, the client can freely engage with the selected items in the room. The therapist monitors the client and logs the item(s) with which the client engages and the duration of the engagement. The items are then placed in order based on total duration. The item with the most extended duration is the most preferred; the item with the shortest duration is the least preferred.

Example
Myra is starting to work with Grover, a 5-year-old boy with autism. Myra is conducting a free operant preference assessment with Grover. Myra takes him to a therapy room that has the following 5 items in the room: a toy car, a toy piano, play-doh, bubbles, and a puzzle. Myra watches Grover engage with the toys for 15 minutes. Myra logs the following data:

Item	**Engagement**
car	1st engagement: 3 minutes 6th engagement: 3 minutes Total: 6 minutes
puzzle	2nd engagement: 2 minutes
Play doh	3rd engagement: 1 minutes
bubbles	4th engagement: 1 minutes 7th engagement: 3 minutes Total: 4 minutes
piano	5th engagement: 2 minutes

Based on this data, the items are ranked as follows from most preferred to least preferred:

1. Car
2. Bubbles
3. Puzzle and piano
4. Play-doh

Single Stimulus Preference Assessment

A single stimulus preference assessment is conducted when a client cannot select between highly preferred and low-preferred items. This assessment is performed at a table or on the floor. To run this assessment, choose the items and present them one at a time. Record the duration of the engagement. Continue this process until all items have been presented. If the client does not engage, mark this a no approach.

Example			
Trial	**Item Name**	**Approach**	**Engagement (mm:ss)**
1	Play-Doh	Yes No	1:53
2	school bus	Yes No	2:21
3	ball	Yes No	0:00
4	toy piano	Yes No	0:45
5	stacking toy	Yes No	1:23

Based on this data, the items are ranked as follows from most preferred to least preferred:

1. School bus
2. Play-doh
3. Stacking toy
4. Piano
5. Ball

Paired Stimulus Preference Assessment

A paired stimulus assessment is conducted at a table or floor area by placing two items in front of the client and asking the client to select an object. When an item is selected, the client is allowed to engage with the item for 10-30 seconds before it is removed. The assessment is discontinued after all items have been paired or when the client does not respond to three consecutive paired stimuli presentations.

Example			
Trial	**Item Selected**	**Item Selection**	**Preference Hierarchy**
1	A B	A selected **0** times	Highest preferred item: **C**
2	C A	B selected **2** times	Moderately preferred items: **B** and **D**
3	A D	C selected **3** times	Lowest preferred item: **A**
4	B C	D selected **1** times	
5	D B		
6	C D		

Multiple Stimulus without Replacement Preference Assessment

A multiple stimulus preference assessment without replacement is conducted at a table or floor area with the client by placing all the times in front of the client. Ask the client to "pick one." Allow the client to engage with the item for 5-30 seconds before removing it. While the client is engaging with the item, move the leftmost item to the rightmost position to assist in detecting bias. Repeat the trials until no items are left.

This process is completed several times with the client to determine the preference of the items.

Example		
Trial	**Item Selected**	**Placement of Item**
1	C	A B C D E
2	A	B D E A
3	B	D E B
4	E	E D
5	D	D

If the trials all resulted in the same selection, the preference is as follows:

1. C
2. A
3. B
4. E
5. D

Multiple Stimulus with Replacement Preference Assessment

Multiple stimuli with replacement preferences assessment is conducted at a table or floor area with the client by placing 3-4 items in front of the client. Prompt the client to "pick one" and allow the client to engage with the selected item for 5-30 seconds before removing the item. While the client engages with the item, replace the unselected items with new ones. Place the selected item back in with the items to be selected. Repeat the process until all items have been presented at least two times, or when the client refuses to make a selection.

The following table provides an overview of multiple stimuli with replacement preference assessment.

<table>
<tr><th colspan="4">Example</th></tr>
<tr><th>Trial</th><th>Placement of Item</th><th>Items Selected</th><th>Preference Hierarchy</th></tr>
<tr><td>1</td><td>A B C</td><td>Item A: 0 times</td><td>Highest preferred item:
B</td></tr>
<tr><td>2</td><td>C D E</td><td>Item B: 2 times</td><td>Moderately preferred item:
C</td></tr>
<tr><td>3</td><td>A B C</td><td>Item C: 3 times</td><td rowspan="3">Lowest preferred items:
A, D, and E</td></tr>
<tr><td>4</td><td>D A B</td><td>Item D: 0 times</td></tr>
<tr><td>5</td><td>C B E</td><td>Item E: 0 times</td></tr>
</table>

Based on this data, the items are ranked as follows from most preferred to least preferred:

1. C
2. B
3. A, D, and E

B-2 Functional Assessment Procedures

Assist with functional assessment procedures

Curriculum-based

Developmental

Social skills

Each client is different. The assessments used for each client differ depending on the individual's needs. A BCBA may use observations, surveys, questionnaires, interviews, and assessment tools to assess an individual. These tools can be curriculum-based, developmental, or social skills based. The type used depends on the individual's needs.

The following table provides an overview of the different types of assessments. Note that some assessments may overlap categories and touch on all three areas.

Assessments	
Type	**Description**
Curriculum-based	• Assessment of taught skills • Can focus on academics such as math, writing, or reading
Developmental	• Focuses on developmental milestones • Focuses on age- and grade-level skills
Social skills	• Focuses on age- and grade-level social skills

B-3 Functional Behavior Assessment

Assist with functional behavior assessment procedures

Indirect

Descriptive

Functional analysis

A functional behavior assessment helps the BCBA determine the function of a behavior based on the antecedent and consequences. There are three types of FBA processes: indirect, direct, and functional analysis (which will be covered later in this section). The results of the indirect and direction assessments include:

- Operationally defined target behaviors
- Antecedents for target behaviors

- The hypothesized function of the target behaviors
- Replacements for target behaviors

Step for Conducting an FBA

The following list provides the 5 steps for conducting an FBA.

1. Identify the problem
2. Collect information to determine the function
3. Form a hypothesis
4. Plan an intervention
5. Evaluate the plan

Indirect Assessment

An indirect assessment is a process in which the client is not directly observed. Instead, the BCBA may use:

- Existing behavioral data
- Checklists
- Rating scales
- Interviews
- Surveys

The BCBA may request this information from the schools, teachers, other therapists, and caregivers. The information is used to determine the context in which the behavior occurs. The indirect assessment information is typically insufficient to formulate a functional hypothesis.

An indirect assessment is conducted to gather information about when and where a challenging behavior occurs and why the behavior occurs.

Direct Assessment

A direct assessment is a process in which the client is observed in the natural environment and recorded data. The data are then used to determine the antecedent for the behavior and hypothesize the function of the behavior. The data are used to develop an operational definition for the target behavior. The following table provides an overview of descriptive measures used when conducting a direct assessment.

<table>
<tr><th>Descriptive Measure</th><th>Description</th><th>Example</th></tr>
<tr><td>ABC event recording</td><td>Allows the BCBA to identify the antecedent and consequence for a behavior. The information can be used to formulate a hypothesis for the function of the behavior.</td><td><table><tr><th>Antecedent</th><th>Behavior</th><th>Consequence</th></tr><tr><td>Mom asked Charlie to clean his room</td><td>Charlie fell to the floor and cried</td><td>Charlie did not have to clean his room</td></tr></table></td></tr>
<tr><td>Antecedent manipulation</td><td>Identifying triggers for target behaviors and altering the environment before a behavior occurs</td><td>Using the ABC data above:
Antecedent manipulation: Mom tells Charlie to first clean his room, then he can play his video game an extra 10 minutes
Behavior: Charlie cleans his room
Consequence: Charlie gets to play his video game an extra 10 minutes.</td></tr>
<tr><td>Functional Analysis</td><td>Called an "FA". This procedure involves manipulating events and recording data to determine the function of a behavior.</td><td>By manipulating events, the BCBA can determine if the behavior is a function of attention, escape, access, or self-stimulation.</td></tr>
</table>

An FA manipulates events to determine the function of a behavior. The following table provides more information on conducting an FA on a fictitious client.

Max is a 10-year-old boy who lives with his mom and dad in San Diego, CA. Max exhibits tantrum behaviors when he hears loud noises. The BCBA working with Max observes his behavior and documents the following:

Antecedent	Behavior	Consequence
Max is in the living room when a loud noise occurs in the kitchen	Max looks toward the kitchen then returns to his game	Max is allowed to continue his game
Max is in the living room with mom when her phone rings	Max looks at mom and exhibits tantrum behaviors	Mom rushes to Max to make sure he is OK
Max is playing in the garage when a box falls on the floor	Max gasps and looks at the box, then continues playing	Max plays for the next 15 minutes.

The BCBA concluded that the behavior might function as attention. A FA is conducted to determine the function of the tantrum behavior.

Conducting an FA				
	Attention	**Escape**	**Sensory**	**Play**
Description	(30 min) play a loud sound and provide attention when Max has a tantrum	(30 min) play a loud sound and allow Max to escape the task he is working on	(30 min) Max sits in the room. Play a loud sound and record if Max becomes upset or has tantrum behaviors	(30 min) Provide Max with a variety of reinforcing activities/items. Document if Max exhibits tantrums when a loud noise occurs while he is playing.
Data over 8 sessions	24 tantrums	15 tantrums	2 tantrums	2 tantrums

The data from the FA are graphed below in an alternating treatment design. Based on the graph, it can be concluded that the function of the tantrum behavior is attention. The tantrum behavior during the Contingent Attention phase occurred more frequently than in the other conditions.

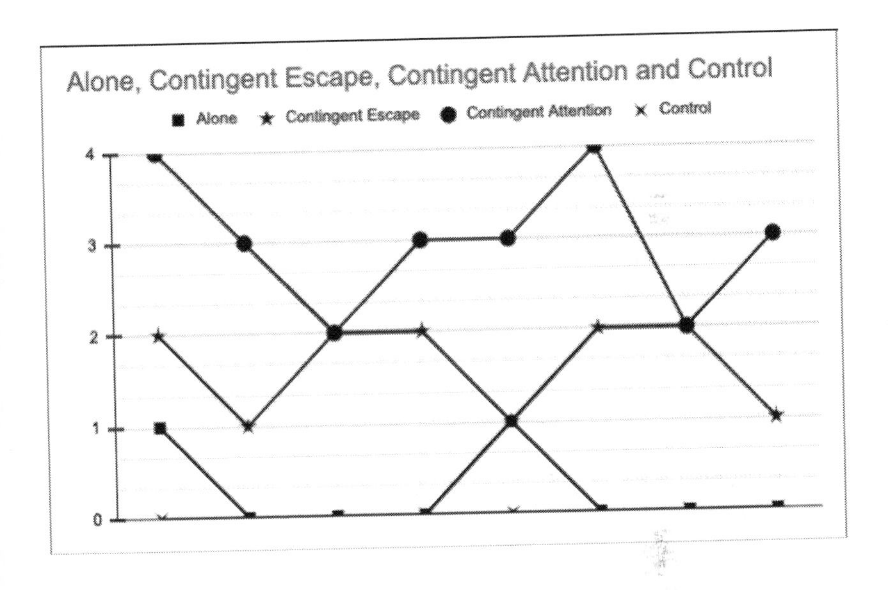

3

Chapter 3

C -Skills Acquisition

C-1 Skill Acquisition Plan

Identify the essential components of a well-written skills acquisition plan.

Skill definition

Baseline measurement

Clear goals

Description of the procedures

Reactive strategies

Data collection and graphing

Plan review

Maintenance and termination

A supervisor or another BCBA writes a skill acquisition plan. You are responsible for understanding and performing the items listed above. The following sections provide more detail on each item.

Skill Acquisition Plan Details	
Skill definition	The definition of the skill acquisition plan is created when the plan is created. It is your responsibility as an RBT to review and understand the plan's definition.
Baseline measurement	You are responsible for recording behaviors before the intervention to determine the present level of the behaviors. This procedure creates a baseline measurement. The behavior during/after intervention can be compared to the behavior before intervention to determine if the intervention is successful.
Clear goals	Goals are critical to a skill acquisition plan. The plan contains measurable goals that are clear and concise. By being measurable, all individuals involved in the client's progress can easily determine when a goal has been met.
Detailed description	Each goal contains a description. The description must be written so that any therapist (RBT, BCcBA, or BCBA) can take over an existing plan and implement it with fidelity. As an RBT, you are responsible for asking questions about descriptions that may be vague, unclear, or missing from the plan.
Reactive strategies	Each plan contains goals that address maladaptive behaviors. The plan includes information on how to punish or extinguish these behaviors and the replacement behavior.
Data collection and graphing	Companies vary in how data are collected. Some companies use paper; others use online programs for data collection. The plan provides information on how the data will be collected, organized, and displayed.

Skill Acquisition Plan Details	
Plan review	The minimum amount of supervision set by the BACB is 5% of the hours spent with clients and at least twice a month. During the supervision visits, the BCBA will review and update the plan.
Maintenance and Termination	When a client has met all goals or it is determined that the client's level of functioning is commensurate with the age, termination can be discussed.

C-2 Session Preparation

Prepare for the sessions as required by the skill acquisition plan.

Read the plan

Review

Client communication

Implement

Preparing for a session can be simple and easy to implement, or it can be complex and require practice to implement. The BCBA creates the plan for you to implement with the client. When a new plan is completed, it is recommended that you:

- Gain access to the plan so that you can read it before seeing the client
- Practice the interventions listed in the plan to ensure you can run them in the session
- Make notes of any areas of the plan about which you have questions.
- Set an appointment with your supervision/BCBA to review the plan and obtain clarification on questions you have about the plan.
- Discuss with the supervisor who/how/when the plan is reviewed

with the caregivers and client

- Make any job aids that may help you run the session.
- Obtain all the materials that are needed to run the plan
- Make sure to include data sheets or devices for data collection.
- Implement the plan during client sessions

If you have questions during the session, make notes of the questions and follow up with your supervisor after the session. Do not wait until the subsequent supervision.

C-3 Contingencies of Reinforcement

Use contingencies of reinforcement

Unconditioned reinforcement
Conditioned reinforcement
Unconditioned punishment
Conditioned punishment
Continuous vs. intermittent reinforcement
Schedules of reinforcement and punishment

Conditioned vs. Unconditioned

Reinforcement and punishment can be conditioned or unconditioned. The difference between conditioned and unconditioned is: unlearned vs. learned. The following table provides more information on the differences between unconditioned reinforcement and punishment and conditioned reinforcement and punishment.

Conditioned vs. Unconditioned	
Unconditioned reinforcement	Reinforcement that is unlearned. These items that are needed to live. For example, water, sleep, and food. An individual does not need to "learn" that these items are reinforcing.
Conditioned reinforcement	Reinforcement that is learned. An individual is not reinforced with a conditioned reinforcer until the item is paired with something that is reinforcing. For example, if someone from a remote village is given a debit card, the card does not function as reinforcement. When the debit card is paired with the ability to gain access to items, the debit card gains reinforcing properties.
Unconditioned punishment	Punishment that is unlearned. For example, if someone touches a hot stove, he/she immediately removes their finger/hand from the stove. The behavior of touching the stove will most likely decrease. Examples of unconditioned punishment include pain, excessive heat or cold, and loud sounds.
Conditioned punishment	Punishment that must be learned The item does not possess punishing qualities until the thing is paired with something that is punishing. For example, you set a unique ringtone on your phone for your mean boss. This way, you always know when your boss calls. Every time your boss calls, it never goes well and ends with her yelling at you..no, matter what you do. After a few calls, your boss's ringtone will be paired with unproductive talks and yelling. When you hear the ringtone, you don't want to pick up the phone because you don't want to hear the boss yell at you. The ringtone is now conditioned punishment.

Continuous vs. Intermittent Reinforcement

Reinforcement can be delivered every time a behavior occurs, or it can be returned for every other response, every three responses, etc...... ...... The difference between continuous and intermittent is when the reinforcement is delivered. The table below provides an overview of these types of reinforcement.

Continuous vs Intermittent Reinforcement	
Continuous reinforcement	Reinforcement is delivered for every occurrence of the behavior
Intermittent reinforcement	Reinforcement that is not given for each instance of the behavior. Intermittent can be delivered on schedules of reinforcement.

Schedules of Reinforcement and Punishment

Two dimensions categorize schedules of reinforcement and punishment:

Fixed vs. Variable - how often the reinforcement or punishment is delivered

- A fixed schedule has a set number of responses before providing reinforcement or punishment.
- A variable schedule has various responses required before reinforcement or punishment is delivered.

Ratio vs. Interval - on what type of schedule is the reinforcement delivered

- The ratio indicates the number of responses needed before providing reinforcement or punishment.
- Interval indicates the time that must pass before reinforcement or punishment is delivered.

The following table overviews the four types of reinforcement and punishment schedules.

Schedules of Reinforcement and Punishment	
Fixed Ratio	• Reinforcement or punishment is delivered on a fixed number of responses • Provides a high steady rate of responding with a post reinforcement pause
Fixed Interval	• Reinforcement is delivered for the first response after the duration of time has passed since the last correct response • Provides an increasing rate of responding at the end of the interval with a post reinforcement pause
Variable Ratio	• A variable number of responses are required before reinforcement is provided • Provides a high, steady rate of responding • The required number of responses varies between responses; however, the average of the response must equal the variable amount
Variable Interval	• A variable length of time is required before reinforcement is provided • Provides a high, steady rate of responding • The required length of time varies between responses; however, the average of the response must equal the variable length of time

Example of Schedules of Reinforcement and Punishment	
Fixed Ratio	You want to increase the number of math problems that a student completes during class. The student is offered 1 minute of recess for every five problems he completes. This is an FR5 schedule Five problems must be completed before reinforcement is provided.
Fixed Interval	A student is completing a reading assignment. The student can read; however, she asks for help sounding out words as a function of attention. The teacher has been directed to help every 5 minutes. This is a fixed interval.
Variable Ratio	While working on homework, your child becomes distracted easily. You offer a piece of candy for an average of every 3 problems. The child receives a piece of candy after 2 problems, 4 problems, and 3 problems. This is a VR3 schedule. The child is reinforced for an average of every 3 responses. The child does not know when the reinforcement will be delivered; therefore, the child will likely complete problems at a consistent rate.
Variable Interval	While working on homework, your child becomes distracted easily. You offer a piece of candy for an average of every 2 minutes. The child receives a piece of candy after 2 minutes, 4 minutes, 1 minutes, and 2 minutes, 1 minute. This is a VI2 schedule. The child is reinforced for at an average of every 2 minutes. The child does not know when the reinforcement will be delivered; therefore, the child will likely complete problems at a consistent rate.

C-4 Discrete Trial Teaching

Implement discrete-trial teaching procedures

A discrete trial is comprised of three components:

Instruction

Response (or lack of response)

Consequence

Discrete Trial Teaching (DTT) is a procedure that:

- It is used for teaching early skills.

- Involves specific procedures

The process for conducting discrete-trial teaching is comprised of the following five parts:

1. The therapist presents a stimulus to invoke a response from the client.
2. The therapist waits a predetermined length of time for the client to respond. Prompting may be needed for the correct response.
3. The therapist reinforces the correct response.
4. The therapist uses a procedure for correcting incorrect responses, as well as extinguishing incorrect responses.
5. The therapist provides an inter-trial between the end of one trial and the beginning of the next trial.

C-5 Naturalistic Teaching Procedures

Implement naturalistic teaching procedures

Using the natural environment to direct client sessions

Naturalistic teaching or natural environment training (NET) is a branch of ABA that uses the natural environment to facilitate learning. A NET setting can be any environment: therapy room, child's home, playground, or grocery store. The child's interests are used to run the session.

When using naturalistic teaching, you must be aware of satiation and deprivation. The following table provides an overview of satiation and deprivation.

Satiation and Deprivation		
Term	**Definition**	**Example**
Satiation	• Being satisfied and reaching full gratification • Value of reinforcers is reduced	The therapist uses the trampoline at the client's home as a reinforcer during the session. One afternoon the therapist rewards the client with the trampoline; however, the client has no desire to jump. When asked, the mom tells the therapist that the child just left a trampoline park with her cousins. The client is satiated, and the trampoline does not have a reinforcing value today.
Deprivation	• Reduced access to an item • Value of reinforcers is increased	The therapist knows the child loves strawberries. The therapist asks the mom not to allow Katie to have strawberries before therapy. The therapist asks the mom to place strawberries in a bowl when she arrives. The child sees the strawberries and uses this opportunity to request items.

C-6 Task Analysis

Implement task analysis training procedures

Break down complex skills into smaller steps

It can be challenging to learn when a skill consists of multiple steps. The skill can be broken down into smaller steps. An individual can learn the small steps of the task and chain the steps together to complete the overall task. To determine the steps, you can:

- Observe someone completing the task and document the steps
- Employ the assistance of a subject-matter expert to determine the

steps in the task

- Complete the task by yourself and document the steps

After the individual steps are learned, chaining links the steps together to complete the complex behavior, the following table provides an overview of chaining procedures that can be used.

Chaining Procedures	
Forward chaining	The steps of a task analysis are taught from the first step to the last step. The first step serves as reinforcement for the next step. The first step is taught until mastery. Then the next step is taught, always beginning with step 1. This type of chaining is best used when the beginning steps are more accessible to complete than the later steps.
Backward chaining	The task steps are taught in order; however, the client will complete the last step independently. When the last step has been mastered, the client is given the last two steps to complete independently. This process continues until the client can complete the entire task independently.
Total Task Chaining	A chaining procedure that uses prompting to complete all steps of the chain. This procedure is a prompt fading procedure.

C-7 Discrimination Training

Use discrimination training procedure.

Stimulus discrimination

Response discrimination

Although only stimulus and response discrimination are listed above, there are two other concepts of generalization and discrimination. The following table provides an overview of all four types of discrimination and generalization.

Chaining Procedures	
Forward chaining	The steps of a task analysis are taught from the first step to the last step. The first step serves as reinforcement for the next step. The first step is taught until mastery. Then the next step is taught, always beginning with step 1. This type of chaining is best used when the beginning steps are more accessible to complete than the later steps.
Backward chaining	The task steps are taught in order; however, the client will complete the last step independently. When the last step has been mastered, the client is given the last two steps to complete independently. This process continues until the client can complete the entire task independently.
Total Task Chaining	A chaining procedure that uses prompting to complete all steps of the chain. This procedure is a prompt fading procedure.

Discrimination and Generalization		
Type	**Description**	**Example**
Response discrimination	Occurs when some responses are reinforced while others are not	Your friend from London has moved to Florida and is very excited to go out and explore. You and your friend are out at dinner. Your friend says she needs to wash her hands - she needs to find the loo. When she walks to the back of the restaurant she sees the sign "restroom" and enters to wash her hands.

C-8 Stimulus Control Transfer

Implement stimulus control transfer procedures

When a behavior evoked by an Sd comes under the control of a different Sd

Stimulus control refers to the "control" a stimulus has over the occurrence or non-occurrence of a behavior. The following scenario provides an example of stimulus control.

Example	Stimulus Control
A group of girls talk and laugh as they walk up the steps. When the first girl opens the library door, all the girls begin to whisper and stop laughing.	The library has stimulus control over the vocal behavior of the girls.

Stimulus control occurs every day in our day-to-day lives. During a session, the therapist may transfer control from the site of cards on the table to the question asked by the therapist.

C-9 Prompt and Prompt Fading

Implement prompt and prompt fading procedures

Response prompts

Stimulus prompts

Prompts are broken into two categories: response prompts, and stimulus prompts. The following table provides an overview of response prompts.

Response Prompts	
Prompt	**Description**
Modeling	A therapist performs the task to provide a visual demonstration of the task. The model provides learners with imitation skills to attempt to perform the task based on what was learned from the demonstration.
Physical guidance	The therapist provides partial or full physical guidance "as needed" to assist with completing the task.
Verbal prompting	The therapist uses spoken words, signs, and pictures to prompt the client to complete the task. This type of prompt is the most used.

When prompting is used, prompt fading must also be implemented to remove prompting and allow the individual to complete the task without assistance. The following table provides an overview of prompt fading techniques.

Prompt Fading	
Prompt	**Description**
Most-to-least	The greatest amount of prompting at the beginning and slowly fade the prompting until the individual can complete the task independently. This type of prompting is associated with errorless learning and the highest completion level.
Least-to-most	The type of prompt fading is used with a fixed time interval. The task is presented to the individual with a fixed time interval to respond. If the correct response is provided during the time interval, reinforcement is provided. If no response or incorrect response is provided, the trial is represented with the least invasive prompt.
Graduated guidance	Graduated guidance is used with physical prompts. When using this type of prompt fading, physical prompting is faded immediately to no physical contact with the client. For example, when prompting to "clap hands", the therapist can physically prompt hand-over-hand and immediately fade to wrists, forearms, and then elbows.
Time delay	This type of prompt fading involves increasing the time between the stimulus presentation and the response prompt. For example, a stimulus is presented with a response prompt. The time delay between the presented stimulus and the response increases to 1 second. Reinforcement is provided for a correct response, and incorrect responses are corrected.

C-10 Generalization and Maintenance

Implement generalization and maintenance procedures

Stimulus generalization

Response generalization

Maintenance

In C-07, stimulus and response generalization were mentioned. More information on generalization is provided in this section.

Generalization is needed for an individual to use what is learned during sessions - in a particular environment or for a particular stimulus - in different environments and settings.

Generalization		
Stimulus generalization	The same response from a stimulus that shares similar features to the original stimulus.	Example: Molly is taught that her pet, a great dane, is a dog. Walking with her mother, she sees a dachshund and says, "dog."
Response generalization	A different, yet similar, response in the presence of a similar stimulus.	When you greet your mom you may say "hi" or "hello". When you greet your friends you may say "What's up" or "Hey".

Maintenance is the last step for a target behavior in an intervention. Maintenance is the continued reinforcement of a behavior and is considered from the beginning. If continued reinforcement for a behavior is not maintained, there is a potential for the behavior to return. Support in the natural environment is necessary for the individual to maintain replacement behaviors. Plan for maintenance by creating goals that will support the natural environment.

C-11 Shaping Procedures

Implement shaping procedures

Shaping

Limitations

Shaping uses differential reinforcement to create new behaviors by reinforcing successive close approximations of the target behavior.

Limitations of shaping are:

- Shaping can be time-consuming if the individual requires many approximations of the target behavior; each instance of the behavior is reinforced. The results from shaping procedures can vary over the implementation of the procedure.
- The therapist must monitor the individual closely to notice slight behavioral changes to reinforce the next approximation.
- The therapist can reinforce unwanted behaviors if the shaping procedure is not implemented correctly.

C-12 Token Economy

Implement a token economy

Characteristics

Implementation

Fading

Token economies are used as a reinforcement procedure. Individuals earn tokens by completing tasks or displaying desired behaviors. The tokens are exchanged for an item/event that the individual wants. To implement a token economy correctly, the therapist must:

- Determine the desired behaviors
- Create a reinforcement schedule for token delivery
- Determine how many tokens are needed to exchange tokens for backup reinforcement
- Provide a backup reinforcer that can be exchanged for the tokens

Token economies consist of two stages: implementation and fading. The following table provides an overview of each step.

Token Economy	
Implementation	• The beginning of the procedure • Tokens are given often as reinforcement (continuous reinforcement) • Tokens are paired with verbal praise
Fading	• The individual knows how the token economy works • The interval between tokens is increased slowly • The number of tokens required for the backup reinforcer is increased • Backup reinforcers are changed to items/activities that occur in the individual's natural environment

4

Chapter 4

D - Behavior Reduction

Behavior reduction

Identify components of a behavior reduction plan

Describe common functions of behavior

Implement interventions based on the modification of antecedents

Implement differential reinforcement procedures

Implement extinction procedures

D-1 Identify the Essential Components of a Behavior Reduction Plan

Behavior reduction plan

Must contain replacement behavior(s)

Uses positive reinforcement

A behavior reduction plan must contain replacement behavior(s) for maladaptive behavior(s). In ABA, positive reinforcement is used to increase the occurrences of replacement behaviors. Punishment is used only if all other options have been exhausted.

D-2 Functions of Behavior

Describe common functions of behavior

Sensory

Escape/avoidance

Attention

Tangible (access)

The function of a behavior is "why" someone does something. All behavior has consequences that either reinforce (increase) or punish (decrease) the behavior. The are four functions of behavior:

1. Sensory
2. Escape
3. Attention
4. Tangible (access)

These functions are used in ABA to determine how a behavior is maintained, and the intervention used to increase or decrease a behavior. The

BCBA uses ABC data to assess the function of a behavior. The following are examples of each function.

Functions of Behavior	
Sensory	• Watching TV because it is enjoyable • Standing in a hot shower because it feels great • Tapping your fingers when you are nervous
Escape	• The teacher puts a worksheet on the desk. The student complains of a stomachache and goes to the clinic. • The parent puts the child to bed, and the child says she is thirsty to get out of bed for a drink. • You take medicine to relieve your headache.
Attention	• A child starts crying when the mom snuggles with her other child. • Jackie bought a new dress to wear for her anniversary for her husband to compliment her on the dress • Your dog runs to the door when you arrive so you can pet her.
Tangible (access)	• The child begs his parents for the PS5 until they buy it for him • The child cries at the checkout counter for candy • You pay your phone bill to have access to call/text your friends

D-3 Interventions based on Antecedents

Implement interventions based on the modification of antecedents such as motivating/establishing operations and discriminative stimulus

Motivating/establishing operations

Discriminative stimulus

To implement an intervention based on motivating/establishing operations and discriminative stimulus, you must first understand the difference between the two.

Motivating operations increase or decrease the value of a reinforcer. Motivating operations can be thought of in terms of satiation and deprivation. If someone is deprived of X, then X is more reinforcing. If someone is satiated with X, then X will be less reinforcing. A discriminative stimulus indicates that reinforcement is, or maybe, available.

Example: You are using cookies as a reinforcer. The cookies are always kept in a blue jar. You sit down to work with your client and place the blue jar on the table. Your client had a big ice cream cone before he came in. When working with your client, he does not seem motivated by the cookies. In this example, the blue jar is a discriminative stimulus that indicates reinforcement of cookies is available. The client had ice cream before he came to the session. The client is satiated; therefore, the cookies are less reinforcing than they would have been if he had not had the ice cream before the session.

These paragraphs can be summarized as follows:

- Motivating operations - depends on deprivation and satiation.
- Discriminative stimuli - a signal that reinforcement is available

D-4 Differential Reinforcement

Implement differential reinforcement procedures

Differential reinforcement of other behaviors (DRO)
Differential reinforcement of alternative behaviors (DRA)
Differential reinforcement of incompatible behaviors (DRI)

There are several differential reinforcement procedures. The procedures targeted for the RBT exam are DRO, DRA, and DRI. The following table provides an overview of each course.

Procedure	Definition	How to Use
DRO Goal is to eliminate a behavior	The target behavior is not reinforced, while **other** behaviors are reinforced on a fixed or variable interval	The client is reinforced when s/he does not engage in the target behavior. The client is reinforced for each interval in which s/he does not call out in class.
DRA Goal is to eliminate a behavior	The target behavior is not reinforced, while the **alternate** replacement behavior is reinforced	The client is reinforced when s/he does not engage in the target behavior. Any other functional alternative behavior is reinforced.

D-5 Extinction Procedures

Implement extinction procedures

- *Extinction*
- *Extinction burst*
- *Resistance to extinction*

Extinction is a procedure in which reinforcement for a behavior is withheld. If the behavior is no longer reinforced, the behavior will decrease over time. To implement an extinction procedure correctly, the function of the behavior must be determined. Once the function is determined, you can appropriately implement an extinction procedure. This procedure may include ignoring behaviors that are attention seeking, eliminating escape behavior by blocking or following through with a demand, removing access to items, or blocking sensory-seeking behaviors.

An extinction burst occurs when the target behavior increases in frequency/duration/intensity after implementing the extinction procedure. An extinction burst means that the behavior will get worse before it gets better. It is common for parents/teachers to stop the procedure when an extinction burst occurs. The parents/teachers do not believe the procedure is "working," while the client is only exhibiting "more" of the behavior to receive the reinforcement that has been withheld for the behavior.

Resistance to extinction occurs when a behavior continues over a period of time when no reinforcement is provided. In resistance to extinction, the behavior occurs at lower rates than before the procedure was implemented and did not have an extinction burst. When intermittent reinforcement, thinned schedules, and variable schedules are used, there may be greater resistance to extinction.

D-6 Crisis/Emergency Procedures

Implement crisis/emergency procedures according to protocol

Crisis/emergency procedures

A crisis/emergency plan is created when a client exhibits self-injurious behaviors or harm to others has been observed. A crisis/emergency plan is discussed with the team and caregivers to ensure everyone knows the tasks involved in the plan and addresses all contingencies. The plan may include physical redirection or restraint, and additional training may be required for the staff. It is essential to review the plan with the team to ensure everyone is aware of the interventions used in the plan.

5

Chapter 5

E - Documentation and Reporting

Documentation and reporting

Effectively communicate

Seek clinical direction

Report on variables

Create objective sessions notes

Comply with all legal requirements for data collection and storage

E-1 - Effectively Communicate

Effectively communicate with a supervisor in an ongoing manner

Meetings

As an RBT, you will meet with your supervisor at least twice each month. Some RBTs may meet with supervisors as much as once per week. The purpose of the meetings is to keep the supervisor aware of the services that are provided, as well as to monitor the behavior of the clients. The supervisor is responsible for creating the plan and ensuring that it is implemented appropriately.

E-2 - Seek Clinical Direction

Actively seek clinical direction from supervisor in a timely manner

Communication as needed

As stated above, an RBT meets with a supervisor at least twice monthly. During these meetings, asks questions and gain support in areas you need to improve. In addition to the meeting, an RBT seeks direction from the supervisor for any topics for which the RBT requires help. The RBT can ask questions using email, text, or phone calls. In addition, if a caregiver asks a question to which you do not have the answer, the question is directed to the BCBA immediately. Do not wait until the next supervision.

E-3 - Report on Variables

Report other variables that might affect the client

Reporting on variables that create a non-productive session

As an RBT, you will collect data on target behaviors. In some cases, other factors may relate to how the client is performing. These factors are documented in your session notes and discussed with the supervisor. Some factors may include medications, locations, or illness.

E-4 - Objective Session Notes

Generate objective session notes for service verification by describing what occurred during the session in accordance with applicable legal, regulatory, and workplace requirements.

Reporting on variables that create a non-productive session

Session notes for a client are written to describe what happened during the session. Letters are written in a manner that is unbiased and objective. A note indicates:

- Where the session was held
- Who was involved in the session
- The activities that occurred during the sessions
- Any relevant information that should be considered when reviewing client session notes

Notes are not written from the RBT's perspective but as an observer of the session and do not include "mentalistic" statements.

The following table provides an example and non-example of session notes.

Session Notes	
Example	**Non-Example**
Tara was seen at home for 2 hours on January 11, 2020. The caregiver participated in the session for the first 30 minutes. Tara worked alone with the therapist for the remaining 1.5 hours.	I saw Tara at her house for 2 hours. Her mom sat at the table with us for 30 minutes but did not help. After her mom left, Tara was a little off and didn't want to do anything.

E-5 - Comply with Applicable Legal Requirements

E-5

Comply with applicable legal, regulatory, and workplace data collection, storage, transportation, and documentation requirements

Retaining client files

The BACB states that all client documentation must be kept for a minimum of 7 years. Some state/local laws may require a more extended retention policy. Please be knowledgeable about the laws for the area in which you work. Also, ensure that client confidentiality is maintained at all times.

6

Chapter 6

F - Professional Conduct and Scope of Practice

Professional conduct and scope of practice

Describe the role of RBT

Respond appropriately to feedback

Communicate with stakeholders

Maintain professional boundaries

Maintain client dignity

F-1 - Describe the Role of RBT

Describe the BACB's RBT supervision requirements and the role of RBTs in the service-delivery system

RBT role

Supervision requirements

Certification Renewals

RBTs are certified paraprofessionals that provide behavior analytic services. RBTs provide direct services, implement behavior interventions and skill acquisition plans, collect and input data, communicate with families as needed, maintain professional relationships, and may assist with functional assessment and individualized assessment procedures. RBTS are not responsible for creating assessment reports or intervention plans.

RBTs must maintain ongoing supervision as per BACB requirements. The following table provides an overview of the BACB requirements.

are effective. Suppose the supervisor observes an RBT and enhances the intervention's effectiveness by implementing a program or responding to a behavior incorrectly. In that case, it is the supervisor's responsibility to ensure feedback is provided to improve the RBT's performance.

For example, suppose a client's tantrum behavior is maintained by attention, and you were to hug the client every time they engage in tantrum behavior. In that case, the supervisor will provide feedback to ensure you are not reinforcing the tantrum behavior.

It is also your responsibility to improve performance accordingly. Behavior analytic services are effective when provided correctly. It is essential to have open communication between you and your supervisor so you can ask questions or make suggestions if applicable. For example, if the BCBA suggests using candy as a reward to improve compliance with a client, but the client's parents do not want their child to have candy, this information is shared with the BCBA. Remember, RBTs do not implement interventions without approval from the supervisor.

F-3 - Communicate with Stakeholders

Communicate with stakeholders (e.g., family, caregivers, and other professionals) as authorized

Stakeholder feedback

When to refer communication to a supervisor

RBTs are on the frontlines of behavior analytic services and are usually the first point of contact for many stakeholders (e.g., teachers and caregivers). It is essential to communicate professionally. Additionally, RBTs do not provide feedback that is beyond their scope of practice. For example, if a caregiver asks specifics about a behavior plan, the caregiver is referred to the supervisor.

RBTs provide session-specific feedback to the caregiver. For example, at the end of the session, the caregiver is informed of how the client is. The feedback includes examples of the client's skills, any maladaptive behaviors, how you managed those behaviors, and any medical or environmental concerns (if applicable). Feedback is brief, objective, and easy to understand. If a client is seen frequently, switch up the feedback so that it is not redundant (parents might feel discouraged if they are given the same feedback). Session feedback ends on a good note.

The following is an example of session-specific feedback:

"Danny had a great session today! We worked on labeling different pictures. He labeled cat, dog, mom, and dad for the first time. He threw a brief tantrum while transitioning to the bathroom, but the demand was followed through. When Danny was inside the bathroom, he stopped crying. He even peed in the potty! I gave him lots of praise and a piece of chocolate. He was laughing and smiling! It was a great day!"

RBTs communicate the following with stakeholders:

- Any medical or environmental concerns that arise during the session (e.g., the client is sick, a sibling is interrupting the session)
- If you need help with something (e.g., client needs a snack, client needs to be changed)
- Session specific feedback

Stakeholders may ask specific questions you cannot answer or change on the client's program, as it is out of your scope of practice. In this case, you should refer the stakeholder to your supervisor. Examples of these scenarios include:

- Questions about scheduling
- Questions about behavior intervention plans
- Insurance questions
- Questions about specific programs

If a stakeholder informs you that they would like you to work on a specific skill during a session that is not in the program, you should refer the concern to the supervisor. For example, if the client's caregiver says that their child cannot tie their shoes, and they would like tying shoes to be added to the program, advise the caregiver that you will inform the BCBA of this concern. If a caregiver tells you that the client started taking a new medication, tell the caregiver that you will let the BCBA know so that it can be added to the client's graphs.

When in doubt, tell the stakeholder you will speak to your supervisor and let your supervisor know of the concern.

F-4 - Maintain Professional Boundaries

Maintain professional boundaries (e.g., avoid dual relationships, conflicts of interest, social media contacts).

Dual relationships and conflicts of interest

Social media

Maintaining professional boundaries is necessary to provide effective behavior analysis services. When boundaries are blurred, caregivers or other stakeholders may take advantage.

A dual relationship is when more than one relationship exists between the RBT and the client outside the role of behavior analysis. Examples include:

- Providing services to your family member or friend
- Providing services to your boss or coworker
- Providing services to your student or teacher
- Providing services to a business associate

We do not enter a therapeutic relationship when there is a dual relationship. However, dual relationships and conflicts of interest can develop. Examples include:

Accepting gifts

- This could constitute a friendly relationship or, in some instances, bribery.

Dating or having sexual relationships with your client

- Providing or accepting professional services

Going to your client's dentist's office for a cleaning

Attending events with the client

- Birthday parties, dinner

Extensive conversations unrelated to session-specific goals

In-depth conversations about personal issues not related to the client.

RBTs should never follow, friend, or communicate with clients on social media. Any of those actions results in a dual relationship.

There are many ways we can prevent dual relationships and conflicts of interest. Think of these as antecedent strategies to maintain professional boundaries. The following list provides a list of antecedent strategies.

Remind your supervisor to share the no gift-giving policy with clients

- Some companies may share a document with clients to inform them of the no-gift-giving policy.
- You may also politely remind your client ahead of holidays that there is a no-gift-giving policy.
- Avoid sharing your birthday or any significant events (e.g., weddings, engagements) to avoid any awkwardness. Weddings may be inevitable, especially if you're taking time off. Remind your clients of the no-gift-giving policy if this is the case.

Do not share your personal email or phone number.

- If your company requires RBTs to be the ones to contact families in the event of last-minute cancellations or reschedules, or vice versa, get a different phone number devoted only to work (e.g., Google voice).

Keep social media accounts private.

Model appropriate conversations, and do not share personal information

F-5 - Maintain Client Dignity

Maintain client dignity

What is client dignity

How to maintain dignity

Client dignity is simple. Show respect for your clients. How do we maintain dignity? Think about your client's needs and respect them. Please do not yell at your clients, don't talk badly about them, don't talk down to them, and understand that they have feelings too. The following list provides some examples of how to maintain client dignity:

Give the client privacy when they are changing or using the bathroom

Allow the client to make their own decisions and choices

- Respect the client's wishes
- For example, if an 18-year-old client vocalizes that they don't want people to know that they have an autism diagnosis, then don't talk about it in front of others.

Share information only with the client and appropriate stakeholders.

- Do not share information without consent.
- Do not post about the client on social media.
- Do not talk about your client with others.
- Do not talk badly about your client in front of others, including the client.
- Consider the risk of harm when implementing interventions.
- Consider the rights of your clients.

There are many ways for us to maintain client dignity. Always ask yourself: Is this fair, is this safe, am I maintaining privacy, and would I be okay if this was happening to me?

Remember that our clients are human with valid feelings. Follow the golden rule: Treat your client as you want to be treated.

The following table provides an overview of the scenarios.

Scenario	**Response**
You're providing ABA therapy for your 18-year-old client. Part of therapy is vocational training, where you attend work with your client. Your job is to stay on the sidelines, and only provide feedback when needed. Your client tells you they like having you there for support, but does not tell anyone, but his boss, why you're there. What do you do?	Respect your client's wishes. If the only stakeholder that needs to know your role is the client's boss, then only the client's boss should know why you're there. Do not wear any clothing that indicates you provide ABA therapy (e.g., no company t-shirts, or cute "Best RBT ever" t-shirts).
You provide services in school for a 9-year-old boy. One of his classmates comes up to you and asks who you are and why you're there. How do you respond?	The classmate is not a stakeholder, so make it up. You can say something like, "I'm helping the teacher," or "I'm a friend." It can be any answer as long as it is appropriate and does not give away who you are.
While at the playground with your client and their mom, your client starts playing with a new peer. This peer's mom introduces herself to you and the client's mom. How do you introduce yourself?	Again, this person is not a stakeholder. Be polite, shake their hand, and say your name. You do not need to tell them why you're there. Sometimes, someone may ask, "Are you their aunt," or "Are you their father." Make it an answer such as "I'm a family friend" or "I'm their cousin."

7

Conclusion

Thank you for prepping with the study guide. The goal of this study guide is to provide easy-to-understand terms and examples to which you can apply the concepts.

There are other study materials available for purchase on the www.behaviorprep.com website. Check out the site for mock exams that simulate the RBT certification exam.

About the Author

I am a Board Certified Behavior Analyst who has worked with individuals with language delays, autism spectrum disorder, anxiety, ADHD, and handwriting difficulties for seven years. Along with professional experience, I also have personal experience with two family members with ASD. I am dedicated to improving the client's and family's quality of life. I received my B.A. degree in Communication Disorders from the University of Florida. I worked as a speech therapist in the school setting before obtaining my M.S. in Exceptional Student Education with a concentration in Applied Behavior Analysis from the University of West Florida. While getting my Master's degree, I was an Instructional Designer and released over ten training manuals. I am now combining my love of ABA, writing, and training to help individuals prepare for their certifications.

You can connect with me on:

https://behaviorprep.com